WHO LIVES ON THE MOON?

Aaron and Melinda Hipple

Paperback ISBN: 978-1-64318-091-5

IMPERIUM PUBLISHING
CREATE YOUR STORY

703 8th Street
Baldwin City, KS, 66006
www.imperiumpublishing.com

This book belongs to

_____

WHO LIVES ON THE MOON?

You, me, the Earth, and the Moon all have a force called "Gravity." The larger something is, the more its gravity can pull other things close to it. The Moon has enough gravity to hold onto cows, but not enough to hold onto air. Cows cannot live on the Moon because there is no air to breathe.

WE HAVE TO BRING OUR OWN AIR SUPPLY.

IT'S COLD UP HERE BUT AT LEAST I CAN BREATH!

It would take six of you on the Moon to weigh the same as one of you on the Earth.

Some people think the Moon is full of holes as if it is made of Swiss cheese. The big circles we see on the surface of the Moon are really called craters. They are left by very large pieces of rock called "meteorites" that fell on the moon millions of years ago. Mice cannot live on the Moon because the Moon really isn't made of cheese.

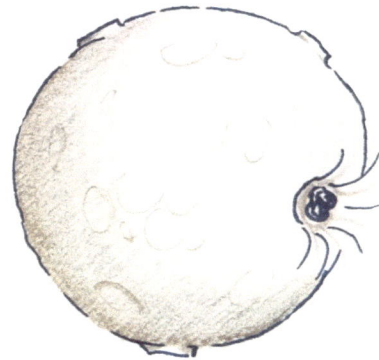

WATCH FOR
FALLING
ROCK

Water is only a little heavier than air. Most water on the Moon floated into space long ago. There is not an ocean, a lake, a river, or even a mudpuddle on the Moon. Whales cannot live on the Moon because there is no water to swim in.

Air and water can escape
the Moon's gravity, but the Earth pulls them close.

Grass needs air and water just like cows and whales do. Grass also needs food called "nutrients." The Moon is covered with rocks and moondust. There are not enough nutrients in the dust for a plant to eat. On Earth, dirt is rich with these elements. Grass and other plants have special roots that dig down to find this food. Grass cannot grow on the Moon becuase there are no nutrients in the soil for it to eat.

There's no food in here. Somebody get me a burger and fries!

BURGER HEAVEN

Do bugs live on the Moon?

The Earth is protected from the very hot Sun because our air bounces many of the hot rays back into space. The Moon has no air to protect it. Some bugs can live in places that are too hot for people, but when the Sun shines directly on the Moon, even a bug becomes a crispy critter. Bugs cannot live on the Moon because the days are too hot.

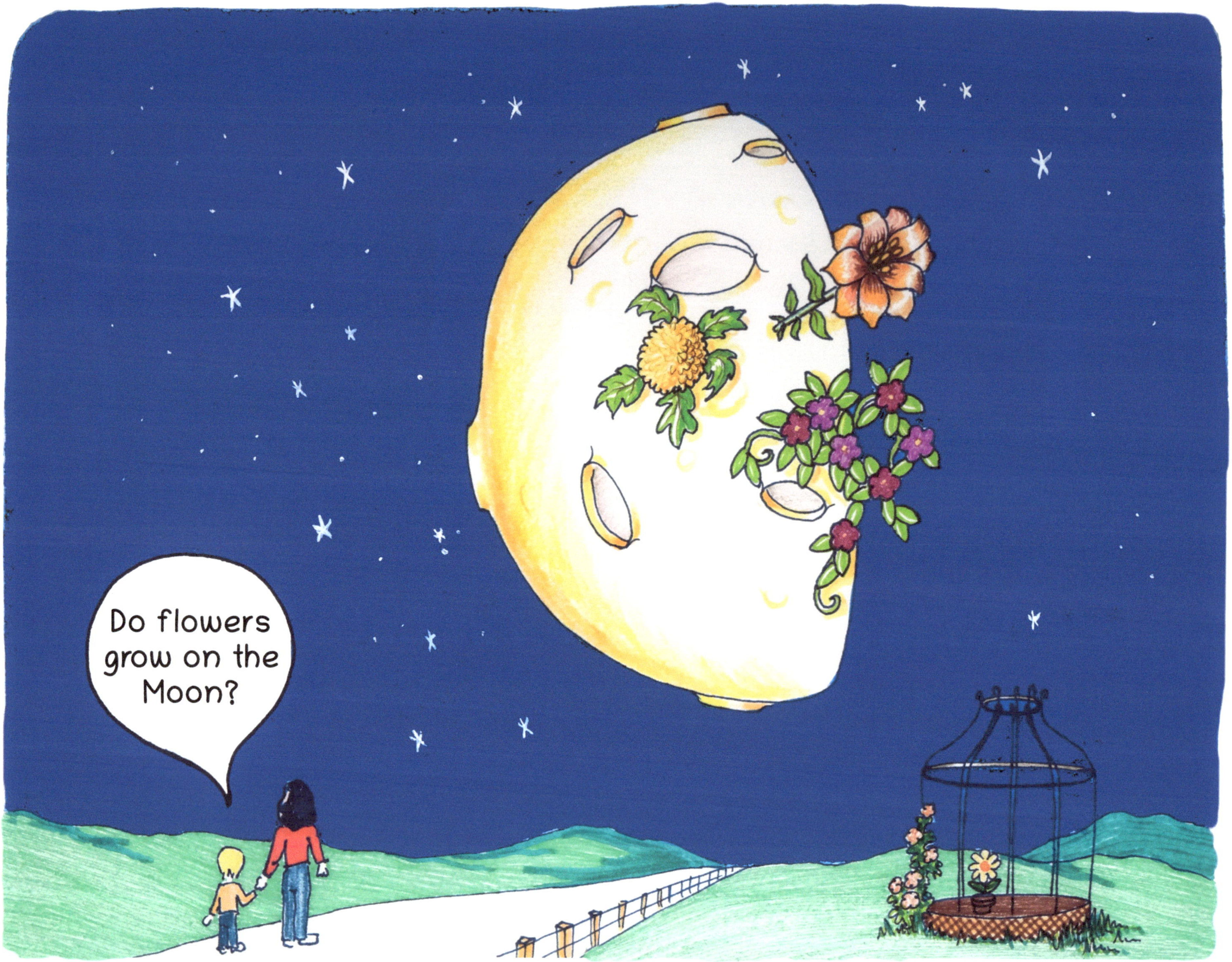
Do flowers grow on the Moon?

The same air that protects us from the hot Sun in the daytime helps hold some heat near the Earth overnight. Without that protective air, we would freeze! On the night side of the Moon, it is so cold that a flower would freeze instantly. The smallest movement could break the flower into hundreds of tiny pieces. Flowers cannot grow on the Moon because the nights are too cold.

When the Earth and Moon were formed long ago, they both were very hot inside. Because the Moon is smaller, it has cooled down. But the Earth still holds heat inside keeping some of its rock melted. When that melted rock pushes up through the crust, it makes a volcano. The Moon does not grow volcanoes because the inside cooled down and turned into a solid ball.

Solid inner core

Moon dust and broken rock

Solid inner core
Melted outer core
Solid mantle
Melted rock
Crust

Volcano

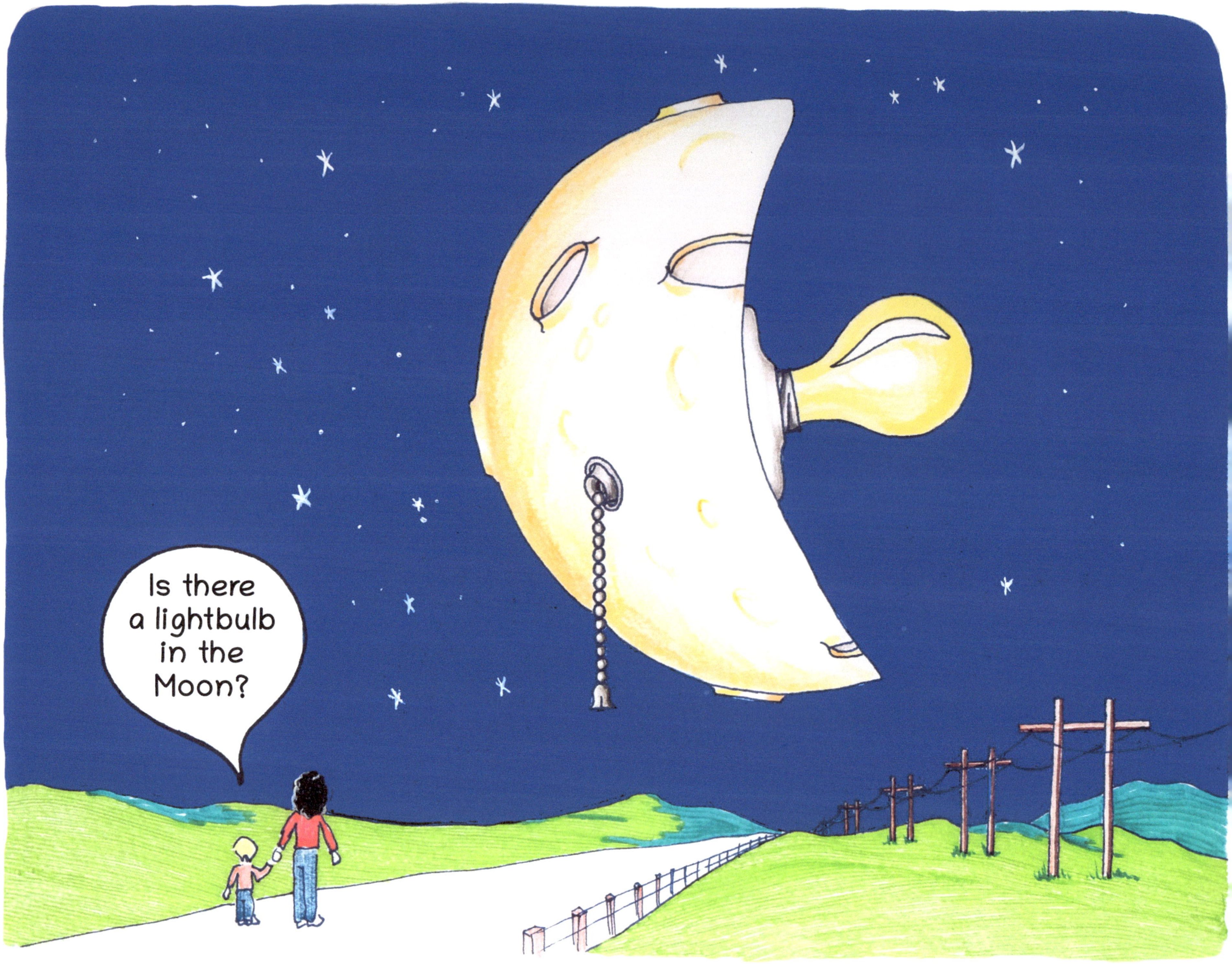
Is there a lightbulb in the Moon?

The Moon is too small to make its own heat or light. The light we see from the Moon is sunrays bouncing off the Moon and back to Earth. If the Moon was big enough to make its own light, it would become another sun. It would be bigger than the Earth!

SUN RAY

Think of it! Two suns!

Yeah! we'd get pretty hot!

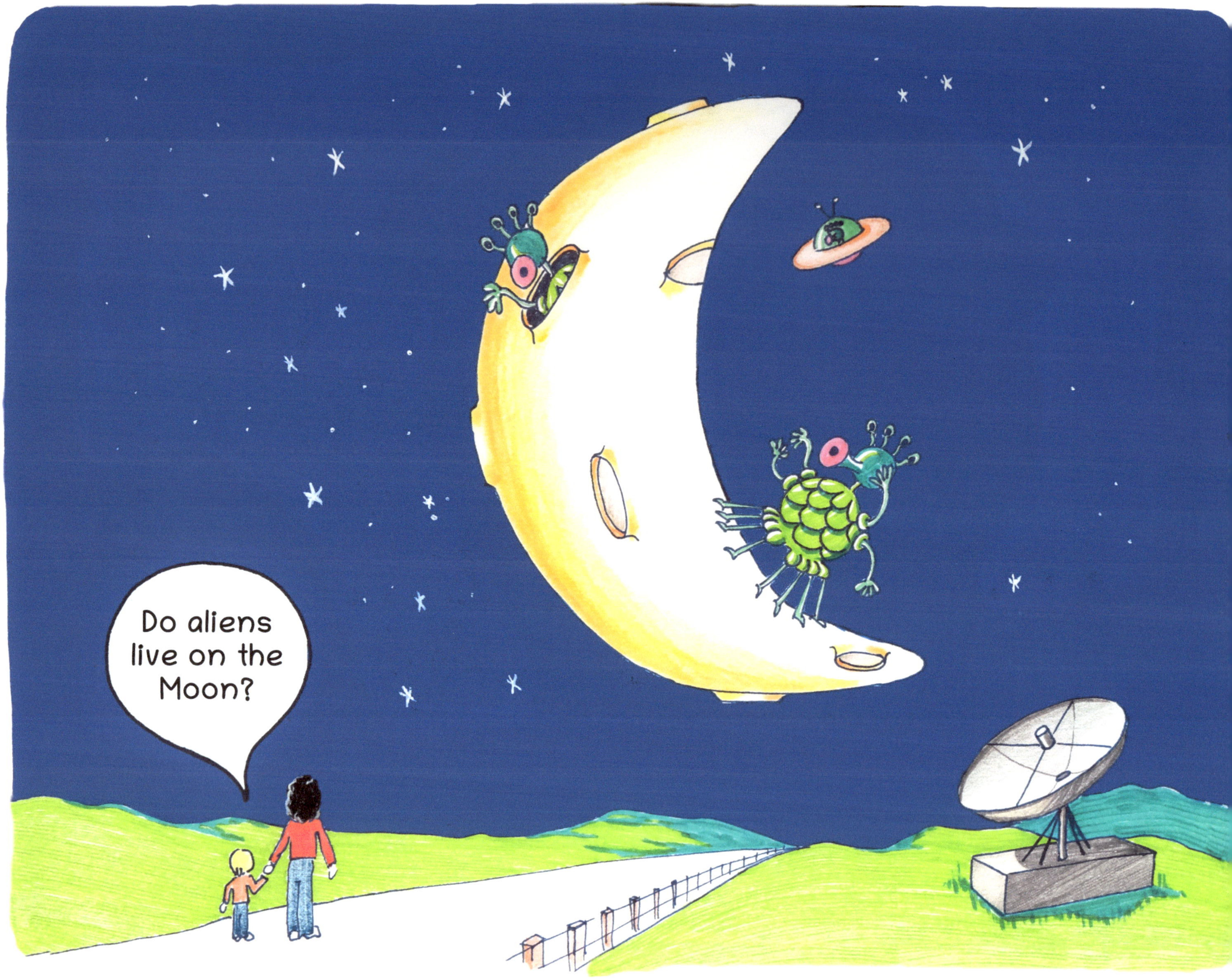
Do aliens live on the Moon?

Scientists have been searching the skies to find signs of life in space. They have sent out messages hoping aliens would hear them and call us back. So far, no aliens have answered our call. We have searched the Moon for signs that aliens have visited. If they ever have, they were very good at cleaning up after themselves.

People have visited the Moon. Perhaps someday we will build a work station there to do scientific experiments and study space. The first people who travel there to live will have to take their own air, water, and food from Earth. Once we have learned to live on the Moon, we can travel to the other planets and beyond.

Perhaps someday you will travel to the Moon on a rocket ship or space shuttle!

www.ingramcontent.com/pod-product-compliance
Lightning Source LLC
Chambersburg PA
CBHW040512230326

41458CB00105B/6510